Grasslands

Philip Steele

 Carolrh **Minneapolis**

All words that appear in **bold** are explained in the glossary that starts on page 30.

Photographs courtesy of: The Hutchison Library 17br / Hilly Janes 25b; Impact Photos / Simon Grosset 7; / Andrew Smith 9cr, 21t; / Phillipi/Visions 10; / David Reed 19b, 20; / Paul Weinberg 21b; / Richard McCaig 23b; / Tom Webster 25t; Peter Newark's American Pictures 11b; South American Pictures / Tony Morrison 14bl, c & r, 15t, 15bl & br; Still Pictures / Muriel Nicolotti - cover; / Daniel Dancer - title page; / Mark Edwards 5t & b; / Roland Seitre 6; / M & C Denis-Huot 9t & 9cl, 22; / Hjalte Tin 17bl, 27; / Stephen Perm 18, 19t; / Sonja Iskow 23t; / Klein/Hubert 24; Zefa 11t, 12, 13t & b, 16, 26.
Illustrations and maps by David Hogg.

This edition first published in the United States in 1996 by Carolrhoda Books, Inc.

A ZOË BOOK

Copyright © 1996 Zoë Books Limited. Originally produced in 1996 by Zoë Books Limited, Winchester, England.

Carolrhoda Books, Inc., c/o The Lerner Group
241 First Avenue North, Minneapolis, MN 55401

Library of Congress Cataloging-in-Publication Data

Steele, Philip.
    Grasslands / by Philip Steele.
        p. cm. — (Geography detective)
    "A Zoë book" — T.p. verso.
    Originally published: Winchester, England: Zoë Books. 1996.
    Includes index.
    Summary: Describes grasslands of the world, including the plants, animals and people found there. Suggests activities and questions for discussion and presents case studies of specific grasslands.
    ISBN 1-57505-042-0 (lib. bdg. : alk. paper)
    1. Grassland ecology — Juvenile literature. 2. Grasslands — Juvenile literature. [1. Grassland ecology. 2. Ecology. 3. Grasslands.]
I. Title. II. Series.
QH541.5.P7S735 1997
574.5'2643 — dc20   96-5710

Printed in Italy by Grafedit SpA.
Bound in the United States of America
1 2 3 4 5 6 02 01 00 99 98 97

Contents

Grasses and Grasslands

There are two main groups of grasses. One is natural, or wild, grasses, which includes about 10,000 different types, or species, of wild plants. The other is cultivated grasses, which farmers grow as food crops. Sometimes called **cereal crops**, cultivated grasses include wheat, oats, barley, corn, and rice.

Natural grasslands are found in most parts of the world. They cover nearly one-fourth of the earth. Before farmers began to cultivate grasslands, natural grasses covered nearly half the earth.

In this book, the term "grasslands" refers to both cultivated and natural grasslands. Forests have also been cleared to raise cultivated grasses, but this book focuses on native grasslands. These areas lie between deserts and forests, in regions that get enough rain to prevent the land from turning into desert, but not enough for forests to grow. These regions fall into two main climate groups — **temperate** grasslands, and **tropical** grasslands, or **savanna**.

● Grass roots form a dense tangle of tiny fibers. One rye plant grew 387 miles of root fibers within a 14-inch cube of soil! End to end, these roots would stretch all the way across the state of South Dakota.

▼ This drawing is a cross section through the temperate grasslands, or **prairies**, of North America. It shows that the height of natural prairie grass varies with the amount of rainfall. Little of this natural grassland is left on the prairies.

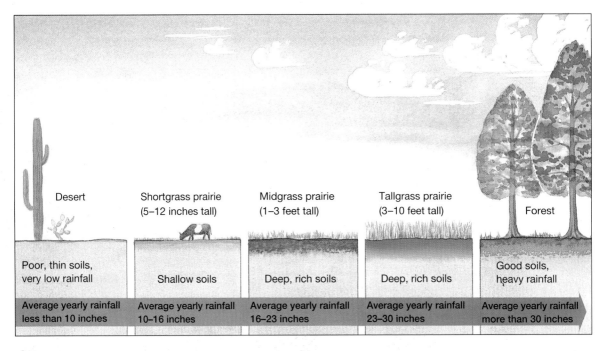

Desert	Shortgrass prairie (5–12 inches tall)	Midgrass prairie (1–3 feet tall)	Tallgrass prairie (3–10 feet tall)	Forest
Poor, thin soils, very low rainfall	Shallow soils	Deep, rich soils	Deep, rich soils	Good soils, heavy rainfall
Average yearly rainfall less than 10 inches	Average yearly rainfall 10–16 inches	Average yearly rainfall 16–23 inches	Average yearly rainfall 23–30 inches	Average yearly rainfall more than 30 inches

◀ Here, farmers are harvesting wheat on the temperate grasslands of Australia. Wheat is a cultivated grass, or cereal. Cereals are grown in many grassland regions.

● Spanish settlers invaded central America about 400 years ago. They borrowed the word for a grassy plain, *zavana*, from the Carib people who lived there. We use the word "savanna" to describe tropical grasslands around the world.

◀ Tropical forests once covered this cultivated grassland in southern India. In many other parts of the world, forests have also been cleared for farming.

Geography Detective

The following foods come from countries around the world. They all make use of cereal crops. Use reference books to find out which cereal is used to make each food. Risotto, spaghetti, tortilla, cornbread, oatmeal, pumpernickel, couscous, chapati.

Find out which cereal crops are grown in the country where you live. Which foods or drinks are they made into?

5

Where are the Grasslands?

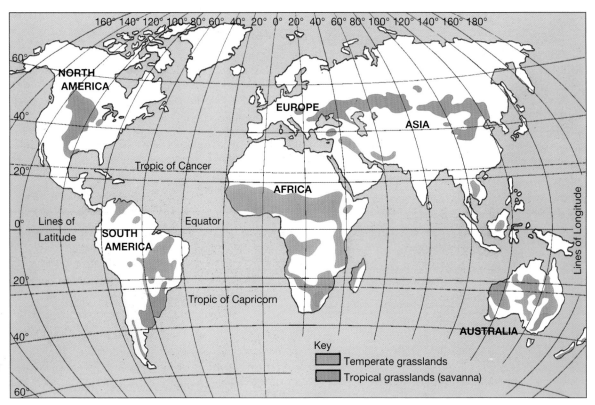

This map shows latitude and longitude lines over the continents, with the Key:
- Temperate grasslands
- Tropical grasslands (savanna)

Labels on the map: 160° 140° 120° 100° 80° 60° 40° 20° 0° 20° 40° 60° 80° 100° 120° 140° 160° 180°; 60°, 40°, 20°, 0°, 20°, 40°, 60°; NORTH AMERICA, EUROPE, ASIA, Tropic of Cancer, AFRICA, Lines of Latitude, SOUTH AMERICA, Equator, Tropic of Capricorn, AUSTRALIA, Lines of Longitude.

Look at an atlas. You will see lines crossing the maps of countries. These lines are called **latitude** and **longitude**. People draw them on maps to help us find where we are.

▲ This map shows the world's main areas of native grassland. Much of the land in these areas is either cultivated or used for grazing cattle and sheep.

◄ Here, a horseman is capturing a capybara, the world's biggest rodent. It lives in South America on a great plain called the llanos, which stretches across Colombia and Venezuela. The llanos includes marshes and forests, as well as wide areas of savanna dotted with palm trees. The plains are dusty and dry for six months of the year. When the Apure and Orinoco Rivers flood each spring, new green grass covers these plains.

◀ These sheep are grazing on North Island, New Zealand. New Zealand has a temperate climate with a high rainfall. The country has some natural grassland, but most of the islands used to be covered with forests. Settlers chopped down the trees and turned the land into **pasture** for grazing sheep and cattle.

● Grasslands are found on every continent except Antarctica. Even there, a few tough grasses can survive. Antarctic hair grass is the southernmost flowering plant recorded on earth. It grows on Refuge Island at a latitude of about 68° south.

Most temperate grasslands lie between latitudes 40° and 60° both north and south of the equator. They are in North and South America, southern Africa, eastern Europe, central Asia, and Australia. The rainfall in these regions averages between 10 and 30 inches a year. Most of the rain falls in late spring and early summer. There may be **droughts**, or dry spells, at other times.

The main regions of hot, tropical grassland, or savanna, lie between latitudes 5° and 20° north and south of the equator. Savanna is found in South America, central and East Africa, and parts of southern Asia and Australia. Here the average yearly rainfall is generally higher than in the temperate lands. It is between 24 and 59 inches. Most of this rain falls during a short wet season. In the dry season there is often no rainfall. This means that drought is common.

Geography Detective

Use an atlas and the map on page 6 to do the following: name one country in eastern Europe and one country in central Asia where there is temperate grassland. Name one country in East Africa and one in South America where there is savanna.

Grassland Wildlife and Soils

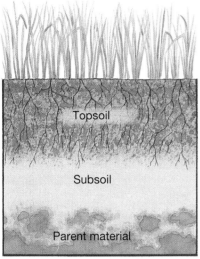

Black prairie soil

Topsoil

Subsoil

Parent material

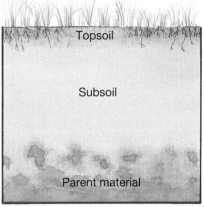

Chestnut-brown prairie soil

Topsoil

Subsoil

Parent material

Very few trees grow on the natural grasslands in temperate regions. The soil is formed over time from rock, or parent material, and decaying plants and animals. Climate also plays a role. In wetter areas, closer to forests, the topsoil is brown and deep. Midgrass prairies, with less rain, have less topsoil but are very rich in **humus**. This layer of decayed plants and animals is very fertile and black. Dry shortgrass prairies have chestnut-colored soil. Savanna soils are mostly red in color.

Many creatures, such as earthworms, live in the soil. The tunnels they create allow air and water to enter the soil, and they help move decaying grasses from the surface into the soil. Insects, birds, and grazing animals also live on the grasslands. Their droppings enrich the soil. Plant-eating animals, or **herbivores**, consume huge amounts of grass. This encourages more grass to grow by exposing new shoots to the sunlight. And animals also help new sprouts by turning up the soil with their hooves and claws. The plants, animals, soils, and climate are all part of the grassland **ecosystem**. Each part of the ecosystem depends on the other parts.

▲ Different areas of grassland contain different types of soil, but most prairie soils are very fertile. Tall and medium-height grasses, including corn, thrive in black prairie soil. Shorter grasses and wheat do well in drier chestnut prairie soil.

● Grasslands are home to many tiny creatures. One square foot may contain 64 earthworms and 1,400 small jumping insects called springtails.

● It takes a long time to digest grass. Cows have 164 feet of intestine and four separate stomachs to do the job!

● When natural grasses die and decay, they add valuable nutrients to the soil. The great variety of plants enriches the soil with many kinds of nutrients.

The light and warmth of the sun helps grass grow. The grass takes food from the soil. Herbivores feed on grass.

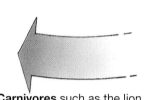

When the lion dies, vultures and other scavengers will eat its body. They take energy from the remains that they feed upon.

Decaying plants and animals and animal droppings enrich the soil and feed the grass. This completes the **energy cycle**.

Herbivores such as the zebra take energy from the grass they eat.

Carnivores such as the lion hunt the zebra. Carnivores take energy from the flesh they eat.

▲ A grassland, like a forest or an ocean, is a system for exchanging life-giving energy. This diagram shows how it works on the East African savanna.

Geography Detective

Many small grassland animals burrow into the ground. Earthworms wriggle through the soil, too. What do you think these creatures do to the soil?

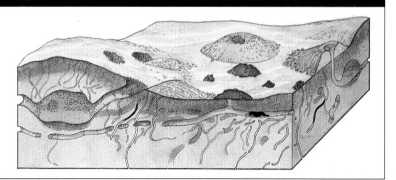

Wind, Dust, and Fire

Most of the world's grasslands are huge, flat plains. Where the natural or cultivated grass grows thickly, its roots stop the soil from blowing away. Winds whip up clouds of soil and dust in dry areas and where farmers plow the land and leave it bare.

On grasslands that are cultivated, farmers must protect the soil from **erosion**, to keep wind and water from carrying the soil away. This is called grassland management. In dry areas, the land may turn into desert if animals graze too much grass. If trees are cut down, there will be no **windbreaks**. On land where only one kind of crop is grown year after year, the soil loses its richness. It becomes loose and dry. When large areas of grassland turn dry and there are no windbreaks, a **dust bowl** can form. Winds blow away the topsoil, darkening the skies.

● In one dust storm in Nebraska in 1934, three million tons of prairie soil blew away.

● In the 1930s, the U.S. government set up the Soil Conservation Service to teach ways of preserving soil. Today many U.S. farmers grow more than one crop in a field and leave stubble on the ground to hold soil in place. Some even allow prairie grasses to grow and enrich the soil when crops have used up the nutrients.

◀ When hot air rises over grasslands, it can lift up the soil into tall spirals called dust devils. Sometimes funnel clouds called tornadoes spin at up to 300 miles an hour. This tornado is raging across Nebraska.

▶ This picture shows flames crackling across the savanna of Botswana in East Africa. Fires are common on grasslands. They spread quickly because the wind fans them. Fires burn trees and scrub, and they also help new grass grow. Farmers sometimes start fires as a way of managing the grassland.

Case Study

Droughts are a natural part of life on grasslands, but during the 1930s the North American prairies experienced a different kind of dry spell. Farmers had plowed up the deep-rooted prairie grasses, which knitted the soil together and held in moisture. In place of grass, the U.S. government paid farmers to plant wheat, which had shallow roots. Each fall, farmers harvested and sold huge quantities of grain. Over the winter, they left the soil bare. Then strong winds whipped across the dry plain. They lifted up the fertile topsoil in great clouds of dust and carried it far away, creating the Dust Bowl.

▼ A dust storm hit this farm in Dallas, South Dakota, in 1936.

Geography Detective

If a dust bowl forms on grassland, what happens to farms and farmers? If crops fail, what happens to people in other parts of the country? Why do people plant grasses and rows of trees and bushes on bare land?

North American Prairies

The temperate grasslands of North America stretch from southern Canada to northern Mexico. Known as the prairies, this huge area was once covered by natural grass. Now, most of it has been plowed up, as have the woodlands that once grew to the east. West of the prairies are mountains and deserts.

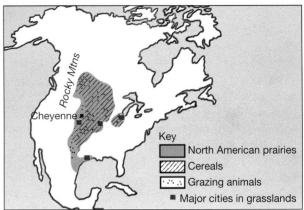

Key

▨ North American prairies
▨ Cereals
▨ Grazing animals
■ Major cities in grasslands

Summers on the prairies are hot and thundery. Winters are cold and snowy. Most rain falls in the east, in states such as Iowa and Missouri. This is the tallgrass prairie, where grass used to grow as tall as an elephant. Today, half the world's corn is grown here.

The central regions have medium-height grasses. This is now a region of wheat farming. To the west lies the Great Plains. This is cowboy country, the cattle-ranching land of the shortgrass prairie. Here, there is less rain than in the east.

Until about 150 years ago, great herds of North American bison, or buffalo, grazed the prairies.

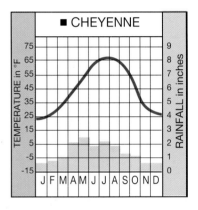

▲ Cheyenne is the capital of the state of Wyoming. The city lies in a cattle-ranching area at the foot of the Rocky Mountains. The rainfall each year is low, averaging 15 inches.

◄ This herd of buffalo is grazing the open prairie in Wind Cave National Park, South Dakota. Some areas of U.S. prairie are now protected. They are in national parks, national grasslands, and wildlife reserves.

▲ These soybeans in Iowa are raised for their seeds and oil. A **legume** plant, soybeans also add nutrients to soil. Farmers plant legumes to enrich soil after growing grain, which uses up nutrients.

Millions of these big herbivores once roamed the plains. Native American peoples hunted buffaloes for hides and meat. Newcomers from Europe, however, killed buffaloes for sport. So many died that the buffalo almost became extinct.

▼ These huge silos, or elevators, are for storing grain. They are in Saskatchewan, a Canadian province. The North American prairies are one of the world's most important **breadbaskets**, or cereal-producing regions.

● Among the Native American peoples who hunted buffalo were the Dakota (called Sioux by European settlers), the Tsistsistas (Cheyenne), and the Absaroka (Crow).

● In 1951, a single field of wheat sown in Alberta, Canada, measured 55 square miles. It is the largest field on record. It was bigger than some small countries!

● The prairie dog is not a dog at all. It is a small rodent that lives in a network of burrows called a "town." One prairie dog town may cover up to 160 acres.

● The term *prairie* comes from a French word meaning "meadow."

Geography Detective

The first farmers on the prairies had to bust through thick sod to plant crops and harvest them using hand tools. Today, huge combine harvesters pick ripe grain. What other inventions, such as machinery or transport, have helped make farming easier on the prairies?

South American Grasslands

South America has both savanna and temperate grasslands. The savanna in Venezuela and Colombia is called the llanos, which means "plains" in Spanish. South of the great rainforests, another savanna covers parts of Brazil and Paraguay. Dry much of the year, both areas of savanna become flooded during the rainy season. Tough axbreaker trees and fan palms dot the grasslands. Cattle ranching is common here.

To the south lie the temperate grasslands called **pampas**, which means "plains" in the Quechua Indian language. These are the flattest grasslands on earth. They cover 190,000 square miles.

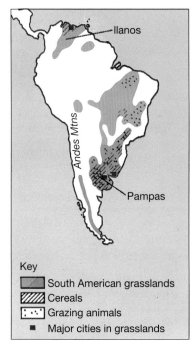

Key
- South American grasslands
- Cereals
- Grazing animals
- Major cities in grasslands

▼ These creatures live on the pampas. The rhea looks like the ostrich of Africa. All over the world, grassland animals live in similar conditions, so many have developed, or evolved, in similar ways.

▲ Rhea

▲ Burrowing owl

▲ Three-banded armadillo

► The job of the gauchos includes rounding up cattle and giving them shots to vaccinate the animals against disease.

● Armadillos live on the pampas. They are the only mammals with armor-plating.

● Some termites live in huge mounds on the South American grasslands. These insects feed on seeds and grass. One of their mounds can contain a million or more termites.

Case Study

The pampas have been farmed or grazed for about 400 years. Each ranch, or estancia, covers a vast area. Cowboys called gauchos round up the cattle herds there. These cowboys are famous for their toughness and for their riding skills.

There are fewer gauchos on the pampas today. Farmers are fencing in the land to grow crops. They feed their cattle on **fodder** such as **alfalfa**. The cattle no longer graze on the open pampas. Four-wheel-drive vehicles, light aircraft, and cattle trucks can reach remote places faster than gauchos on horses.

Geography Detective

Compare these two photographs of South American grasslands. Which one do you think is savanna? Which one is pampa? What are the main differences in the landscape? Which landscape receives more rainfall?

Eurasian Steppes

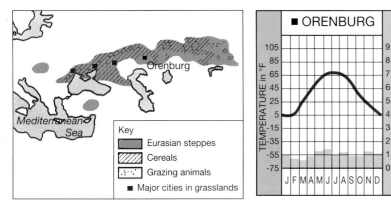

ORENBURG
(temperature and rainfall graph, months J F M A M J J A S O N D)

◀ Orenburg lies on the steppes of northern Kazakhstan, south of the Russian border. Winters are very cold here, but summers are hot.

The Russian word for a treeless plain is **steppe**. The steppes refer to the temperate grasslands of Europe and Asia. They stretch from southern Ukraine through Russia to Kazakhstan and beyond. To the north, scattered trees give way to the great belt of northern forest. To the south are seas and lakes, while to the southeast lie harsh deserts. Wide rivers, such as the Dnieper, the Don, and the Volga, cross the steppes.

The rich black soil of Ukraine and Russia gives way to thinner brown and gray soils on the desert borders. Grain crops grow in large areas of the steppes. Poorer soils provide a little grazing for sheep, goats, and cattle.

● Russia is a big producer of wheat and barley. However, when drought hits the steppes, even Russia has to buy in, or **import**, grain.

● Bustards are the world's heaviest flying birds. At 33 pounds, they can barely take off! The great bustard lives on the Eurasian steppes. Other species live on the grasslands of Africa and Australia.

Key
- Eurasian steppes
- Cereals
- Grazing animals
- ■ Major cities in grasslands

◀ The saiga antelope looks somewhat like a sheep. The strangely shaped nose of this herbivore helps it breathe. Hunting and farming threaten the survival of the saiga antelope.

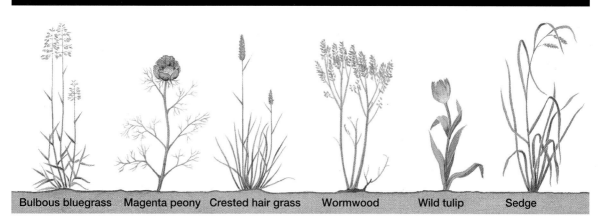

Bulbous bluegrass   Magenta peony   Crested hair grass   Wormwood   Wild tulip   Sedge

▲ Many different species of grass and herbs grow on the steppes. Legume plants such as peas, beans, and clover enrich the soil. In spring these grasslands burst into color as wildflowers bloom.

Many areas of steppe are natural seas of grass. Burrowing rodents, such as marmots, hamsters, and steppe lemmings, live here. They are hunted by the marbled polecat and the ground-nesting eagle.

People have lived on the windy steppes for many thousands of years. They first tamed wild horses here more than 6,000 years ago. Today's steppe dwellers include Ukrainian and Russian farmers and the Cossacks, who are known for horse riding and for lively dances. The peoples of the desert borders include the Kazakhs and the Turkmen. They are farmers and herders. Some of these people still live as **nomads**.

Geography Detective

Look at the two photographs. Which area do you think has richer soil, Kazakhstan or Ukraine? Why did the Kazakh herders have to move with their herds? Why did the Ukrainians settle in villages and farms? Today, many Kazakhs are no longer nomads. Can you think of any disadvantages of living as a nomad? What do nomads live in? How do they keep warm in winter?

Harvesting wheat, Ukraine

Herding sheep, Kazakhstan

Mongolian Steppes

Treeless grasslands grow around the edge of the Gobi Desert in the heart of Asia. They lie close to the border between the country of Mongolia and the region of China called Inner Mongolia. The climate of these steppes is bitterly cold and snowy in winter. In the warm summers, the grass is green and bright with wildflowers. Droughts often occur in the southern parts of these grasslands.

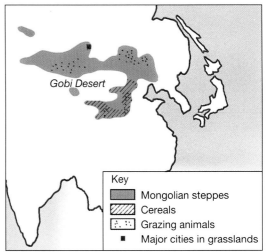

Gobi Desert

Key

▨ Mongolian steppes
▨ Cereals
▨ Grazing animals
■ Major cities in grasslands

Mongolia is one of the world's loneliest countries. Most of it is desert. Only two million people live in an area of 604,250 square miles. Inner Mongolia is smaller, but ten times more people live there. It is more crowded because there are more farms growing cereal crops, more towns and factories. Many people have moved here from other parts of China.

● The country of Mongolia has only 621 miles of paved roads (the United States has 50 times as many miles of road per person). Most people travel on horseback. Others ride on public trucks or four-wheel-drive vehicles.

Case Study

The people of the Mongolian steppes are called Mongols. About 800 years ago, the Mongols

◀ Men, women, and children help round up sheep on the Mongolian grasslands. Mongolian nomads move from one pasture to another with their herds.

These Mongols have set up camp for the summer in the western highlands. They are herding yaks, a tough, shaggy-haired cattle of the Asian mountains. Many Mongols still live in the traditional round tent called a yurt. The tent is made of thick, woolen felt to keep out the winter winds.

Geography Detective

The Mongols live by raising sheep, cattle, goats, and horses. These animals are useful to people. Can you think of some foods that these animals provide? Name 10 things that are made from wool or felt. Name five things made from leather.

conquered lands in Asia and Europe. Their leaders, such as Genghis Khan, were strong, fierce warriors. The Mongols have kept many of their traditions. They love archery, horse racing, and wrestling. Many Mongols still live as nomads. They herd horses, goats, camels, and sheep. Some Mongols have settled in towns. Tourists now visit the grasslands and this brings money to the region.

This beautiful horse is called Przewalski's wild horse, after the explorer Nikolai Przewalski, who first described it to Europeans. The horse stands only 53 inches high at the shoulder. Now extinct in the wild, Przewalski's horse does survive in captivity. Herds are being bred in Europe to be returned to the wild. Another horse of the Mongolian steppes, called the tarpan, became extinct during the 1870s.

Southern African Veld

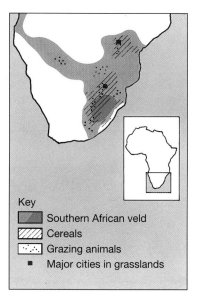

Key
- ▨ Southern African veld
- ▨ Cereals
- ⠄⠄ Grazing animals
- ■ Major cities in grasslands

◀ Savanna covers 90 percent of the land of Botswana. The government has encouraged cattle ranching on the savanna. The meat is sold as an **export** to other countries. Botswana needs the money it earns from exports because it is a very poor country. But the savanna grasses are very thin. The cattle can easily overgraze and destroy the pasture.

Large areas of southern Africa are grasslands. In the north is the savanna of Botswana. It is a hot and dusty region bordering the Kalahari Desert. Farther south is the more temperate grassland of South Africa, Swaziland, and Lesotho. About 300 years ago, Dutch European settlers named this grassland the **veld**, from the Dutch word for "field." The people who are descended from the first European settlers are called Afrikaners.

The grasslands that are between 4,000 and 6,000 feet above sea level are called high veld. Here, the climate is cool and moist. The first European farmers did not manage these grasslands well. As a result, rain washed away much of the fertile soil and crops would not grow. Erosion of the thin, powdery soils in this region is still a problem.

● The European settlers of the 1800s hunted the wildlife on the veld for sport. They made some species extinct. The quagga was a type of zebra with a brown and white body. It was killed for meat and for its tough hide. The last quagga died in a European zoo in 1883.

◀ Giraffes graze fresh green grass on the savanna of Zimbabwe. Huge herds of wild animals used to roam across the veld. As farms and towns spread, the animals had less land to graze. Most of them now live on game reserves like this one. It is Whange National Park in Zimbabwe.

In the drier lowlands, or bush veld, trees and shrubs grow on the grasslands. They are rich in wildlife, including antelope such as the blesbok and the rhebok. Part of the area is in Kruger National Park. This is one of South Africa's **conservation areas**.

Many different peoples live on the southern African grasslands, including the Tswana, the Zulus, and the Swazi. Some of them live by herding or by growing crops. Most of the large farms are still owned by Afrikaners.

Geography Detective

People first hunted animals such as the quagga on foot with spears or arrows. Then the Europeans arrived on the veld with their horses and guns. Why do you think so much wildlife became extinct in the 1800s? How did farming the veld harm its wildlife? Are there any animals in danger today where you live? Find out what is being done to protect them.

◀ A farmer keeps warm on the high veld in South Africa's Transkei region. The climate of these mountain grasslands is cool and damp.

Central African Savanna

Savanna covers a wide belt of land between the Sahara Desert and the dense rainforests of central Africa. It also covers the great plains of East Africa. In parts of Kenya and Tanzania, the savanna is at a height of 3,200 to 6,500 feet. High, snowcapped mountains, such as Mount Kilimanjaro, rise above these flat highlands.

Tough grasses grow on the savanna. The landscape is green during the rainy seasons and brown and dusty during the dry seasons. Flat-topped acacia trees and patches of thorny scrub grow here and there. Strange-looking baobab trees survive the droughts by storing water in their thick trunks. Herds of elephants, giraffes, Cape buffaloes, and wildebeests cross the savanna. Lions hunt the antelope and the zebra.

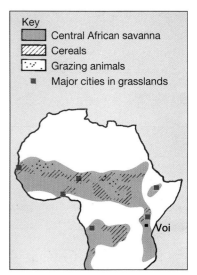

Key
- Central African savanna
- Cereals
- Grazing animals
- ■ Major cities in grasslands

▼ An elephant crosses the dry savanna of Amboseli, in southern Kenya. In the background, Mount Kilimanjaro rises on the Tanzanian border. The East African savanna is one of the last places on earth where large herds of elephants still roam.

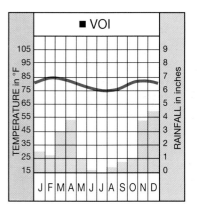

▲ Voi is a town in southern Kenya. The driest months there are from June to September. Farmers grow corn and raise cattle on the nearby savanna. The dry, dusty regions to the north and west are protected within the Tsavo National Park. This is the home of many elephants.

▶ Many Masai people have kept their traditional way of life, but they use modern methods of protecting their cattle from disease.

Some of the Masai lands are now part of national game reserves. The Masai can no longer graze their cattle in these areas.

● The African elephant is the world's largest land animal. A grown male can weigh more than 12,000 pounds. Grass makes up about half its diet.

● Many birds on the savanna use grass to make their nests. Weaver birds' nests are shaped like bags. Hundreds of these beautiful nests hang from acacia trees.

Case Study

The Masai people traditionally live in the area of the East African savanna called the Masai steppe. Part of the Masai steppe is in Kenya, but most of it is in Tanzania. Masai villages are small with thorn fences, or stockades, around them. Cattle herders, the Masai lead their livestock far from the villages in search of pasture during the dry season. Young men called *moran* guard the herds.

Geography Detective

Tourists come to countries such as Kenya and Tanzania to see the wildlife of the savanna. What good and what bad effects might tourism have on the animals? How might it harm the environment? How might it affect the farmers and herders? Hotels and tourist villages need a lot of water. How else might that water be used?

Australian Grasslands

Australia has regions of rainforest, woodland, and scrubland. More than half of the land area is desert and scrub. A very hot desert lies in the center of the continent. These desert wastelands are bordered by grassland. In the north and west, the grasslands form tropical savanna. In the southeast the grasslands are temperate.

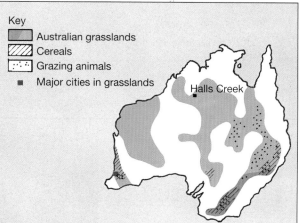

Key
- Australian grasslands
- Cereals
- Grazing animals
- ■ Major cities in grasslands

Halls Creek

For many thousands of years the native peoples of Australia, called Aborigines, lived on the grasslands. They hunted the wild animals there. The region is still rich in wildlife such as leaping kangaroos, burrowing wombats, and large, flightless birds called emus. Some of these animals do not live anywhere else on earth.

European settlers arrived in the 1800s. They brought in sheep, cattle, and seed for cereal farming. Today, Australia produces wheat and barley and raises huge herds of sheep and cattle.

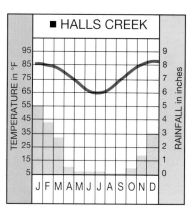

■ HALLS CREEK

TEMPERATURE in °F — 95, 85, 75, 65, 55, 45, 35, 25, 15, 5

RAINFALL in inches — 9, 8, 7, 6, 5, 4, 3, 2, 1, 0

J F M A M J J A S O N D

▲ Halls Creek is a town in the tropical grasslands of northern Australia. The temperature is very high all year. The dry season is from May to September, and the wet season lasts from December to March.

◀ Kangaroos live in the wild in Australia. The female red kangaroo carries her baby, called a joey, in a special pouch. The joey grows in the pouch until it is about eight months old. Pouched animals are called **marsupials**. Many marsupials live on the Australian grasslands.

◀ There are about 10 sheep to every human being in Australia. This picture shows wool being sorted during sheep shearing on a station in New South Wales. Australia is the world's leading producer of wool.

● Australian cattle and sheep farms, called stations, are often huge. One farm covers 11,430 square miles. That is almost as big as all of Florida!

Grassland management is very important in Australia. The grass is often thin in the driest areas, where drought is common. Australia had no large herbivores before Europeans brought in cattle and sheep. These animals can easily overgraze the grasses, and their sharp hooves kick up the light soil. The risk of destroying the pastures is high.

Geography Detective

Australia's first European settlers brought a few rabbits to Australia. The animals breed very quickly and soon they were swarming across the grasslands. What do wild rabbits eat? Why was this a problem for farmers? How do you think they tried to stop the spread of rabbits?

◀ Many huge Australian farms are in regions known as the **outback**. The outback is a long way from any towns or cities. The Flying Doctor Service uses small aircraft to reach remote areas. Doctors provide clinics and an emergency service.

The Future of Grasslands

Grasslands play a vital part in the life of our planet. They produce food for millions of people, and they are home to many kinds of wild animals and plants.

As this book shows, grasslands can easily be destroyed by human actions and by wind, rain, or drought. Huge areas of the world's grasslands are turning into useless desert. This is called **desertification**. It affects two out of every three countries in the world, both rich and poor.

In the United States, soil erosion has turned more than 150,000 square miles into desert. Africa, Asia, and Australia, too, have lost valuable grassland due to poor management of the land.

The grassland areas in eastern Africa are some of the worst hit. Crops have failed and animals have starved to death for lack of food and water. Millions of people have died of **famine**.

▲ Buffaloes like these once roamed wild on the North American prairies. The herds now live on ranches and are raised for their meat. Other wild animals, such as the oryx in East Africa, are also now kept on ranches. The oryx can survive dry conditions better than cattle.

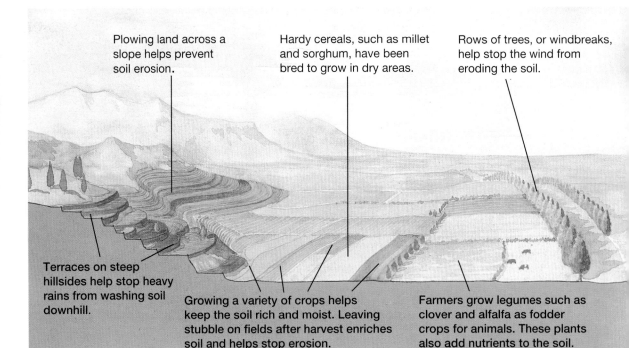

Plowing land across a slope helps prevent soil erosion.

Hardy cereals, such as millet and sorghum, have been bred to grow in dry areas.

Rows of trees, or windbreaks, help stop the wind from eroding the soil.

Terraces on steep hillsides help stop heavy rains from washing soil downhill.

Growing a variety of crops helps keep the soil rich and moist. Leaving stubble on fields after harvest enriches soil and helps stop erosion.

Farmers grow legumes such as clover and alfalfa as fodder crops for animals. These plants also add nutrients to the soil.

● In Sudan, in eastern Africa, the desert has spread by nearly 62 miles. This is a result of drought combined with poor use of the land.

● Poor farming methods on the grasslands result in billions of tons of soil being blown or washed away. The Republic of South Africa lost 440 million tons of topsoil between 1984 and 1994.

▲ Much of Ethiopia is grassland. Famine has struck the country many times. When this happens, other countries send food to help the starving people.

What can be done to save the grasslands? It is a very difficult problem to solve, especially in poor countries where more and more people are trying to survive in dry areas with poor soil. The diagram below shows some ways of managing the land.

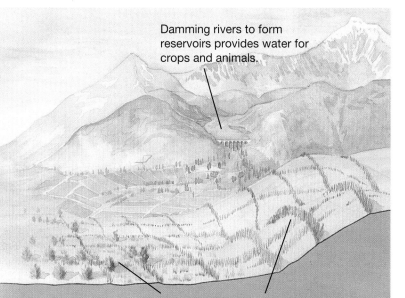

Damming rivers to form reservoirs provides water for crops and animals.

Farmers plant bushes and grasses on nearby sand dunes to stop sand from blowing onto cropland.

Geography Detective

Design and draw a poster called "Save the Grasslands." Use what you have learned in this book to help you.

◄ This is a diagram of an imaginary grassland landscape. It shows some ways in which land can be managed to prevent desertification.

Mapwork

The map on the right shows an imaginary grassland region with a temperate climate. It lies between a northern forest belt and a cool desert in the south. The Crystal Mountain Range lies to the west. The land is divided into four states. Copy the map onto a sheet of paper.

1. Which place would receive more rainfall, place A or place B?

2. Color in yellow the area of shortgrass prairie on the map. Add the color to the map key to show what it means.

3. Can you see two ways of transporting cattle from Camp James to Karlsburg?

4. Which town would have the richer soil, Camp James or Burgundy?

5. a) Choose a picture, or symbol, to show wheat. Draw this on the map in the areas where farmers might grow wheat. Add the symbol to the key.
   b) Choose a symbol to show cattle ranching. Draw this on the map to show the cattle-ranching areas. Add the symbol to the key.

6. A tributary joins the Buffalo River near Camp James. A water pipeline will be built from the river at this point to one of the other towns on the map. Which town do you think needs the water most? Use a dotted blue line to show the route the pipeline might take.

7. Choose a spot on the map to add a national park and draw in the boundaries. Name the park after an animal that you think would live in this region.

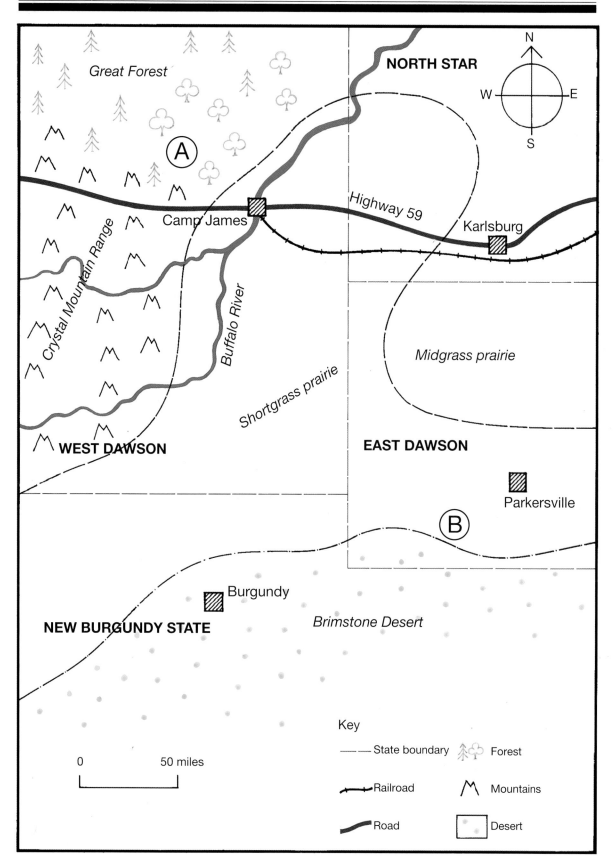

Great Forest

NORTH STAR

N
W — E
S

(A)

Crystal Mountain Range

Highway 59

Camp James

Karlsburg

Buffalo River

Shortgrass prairie

Midgrass prairie

WEST DAWSON

EAST DAWSON

(B)

Parkersville

Burgundy

NEW BURGUNDY STATE

Brimstone Desert

0   50 miles

Key

— — State boundary   Forest

Railroad   Mountains

Road   Desert

Glossary

alfalfa: A plant of the legume family grown to feed cattle.

breadbasket: A major cereal-producing region.

carnivore: An animal that eats flesh.

cereal crop: Any of the grasses grown as food crops, such as wheat, oats, barley, rye, rice, corn, sorghum, or millet. Cereal is also used to describe breakfast foods made from these crops.

conservation area: An area that is carefully protected to prevent misuse or destruction of the natural ecosystem.

desertification: The process of turning fertile land into desert. This can be the result of human activities as well as drought.

drought: A long period of extreme dryness due to lack of rain or snow.

dust bowl: A grassland region that suffers from long droughts and severe soil erosion usually due to overgrazing or overfarming.

ecosystem: A complex community of living and nonliving things that exist as a balanced unit in nature.

energy cycle: The movement of food and energy through an ecosystem, starting with the sun and continuing through a chain of parts in the system. The cycle repeats over and over.

erosion: The wearing away of the earth's surface by wind, water, or ice.

export: A trade term for a good sold from one country to another. Cereals grown on grasslands are exported around the world.

famine: A severe shortage of food, leading to widespread starvation.

fodder: Food, such as hay or silage, fed to livestock (farm animals).

herbivore: An animal that eats plants.

humus: A rich, dark layer of soil formed from decayed plants and animals.

import: A trade term meaning to buy goods from another country for use in your own country.

latitude: Lines that mapmakers draw on maps and globes to divide up the surface of the earth. Lines of latitude circle the earth. They measure, in degrees, the distance north or south from the equator, which circles the globe at 0° latitude.

legume: A large group of herbs, shrubs, and trees that belong to the pea family. The plants add rich nutrients to the soil. The seeds of many legumes, including lentils, chickpeas, and pinto beans, are used for food. Other legumes, such as clover and alfalfa, provide pasture for grazing animals.

longitude: Lines on maps and globes that run from the North Pole to the South Pole. Lines of longitude are measured in degrees and lie either to the east or west of 0° longitude, which runs through London, England.

marsupial: An animal whose young develop outside the mother's body, in a pouch. Kangaroos, wombats, and opossums are marsupials.

nomad: A member of a people that moves their homes with the change in seasons in order to find food for themselves or pasture for their animals.

outback: Remote grassland or desert in Australia.

pampa: The South American term for temperate grassland.

pasture: Grassland or fields used for grazing by animals.

prairie: The North American term for temperate grassland.

savanna: A tropical grassland with scattered trees.

steppe: The temperate grassland of Europe and Asia. The term is also sometimes used to describe shortgrass regions on other continents.

temperate: Falling within the middle latitudes on the globe and having a mild climate with cool winters and warm summers.

tropical: Located near the equator and having a hot climate year-round.

veld: The term used for grassland in southern Africa.

windbreak: A row of trees or shrubs or a fence that breaks the force of the wind on an open plain.

METRIC CONVERSION CHART		
WHEN YOU KNOW	**MULTIPLY BY**	**TO FIND**
inches	25.4	millimeters
inches	2.54	centimeters
feet	0.3048	meters
miles	1.609	kilometers
square miles	2.59	square kilometers
acres	0.4047	hectares
gallons	3.78	liters
degrees Fahrenheit	.56 (after subtracting 32)	degrees Celsius

Index